Crocker Land Expedition

TO THE

North Polar Regions

(GEORGE BORUP MEMORIAL)

1913-1915

D. B. MacMillan, Photo.

S. S. "DIANA"

Newfoundland sealer of 475 tons register, chartered to take the CROCKER LAND EXPEDITION northward, leaving Brooklyn Navy Yard, New York, about 2 July, 1913. She is famous for the part that she has already taken in arctic explorations.

Crocker Land Expedition

TO THE

North Polar Regions

(GEORGE BORUP MEMORIAL)

Under the auspices of the

AMERICAN MUSEUM OF NATURAL HISTORY

AND THE

AMERICAN GEOGRAPHICAL SOCIETY

with the co-operation of the

UNIVERSITY OF ILLINOIS

and the assistance of other Institutions, Federal Departments and Individuals

1913 - 1915

ORGANIZING COMMITTEES

OF THE

CROCKER LAND EXPEDITION

HENRY FAIRFIELD OSBORN, President, American Museum of Natural History
CHANDLER ROBBINS, Chairman of Council, American Geographical Society
THOMAS H. HUBBARD, President, Peary Arctic Club
WALTER B. JAMES, Vice-President, American Geographical Society
EDMUND J. JAMES, President, University of Illinois

EDMUND OTIS HOVEY, American Museum of Natural History
HERBERT L. BRIDGMAN, Peary Arctic Club
WILLIAM S. BAYLEY, University of Illinois

Scientific Staff

DONALD B. MacMILLAN, A. B., A. M., F. R. G. S., Leader and Anthropologist
W. ELMER EKBLAW, A. B., A. M., Geologist and Botanist
FITZHUGH GREEN, U. S. N., Engineer and Physicist
MAURICE C. TANQUARY, A. B., A. M., Ph. D., Zoölogist
——————— Surgeon

West 77th Street and Central Park West
New York, 12 May, 1913

CROCKER LAND EXPEDITION

(George Borup Memorial)

FOREWORD

A YEAR ago the Crocker Land Expedition under the leadership of George Borup and Donald B. MacMillan was far advanced toward complete equipment, and everything looked promising for a successful start in July, 1912. The enthusiasm of the leaders and others associated in the enterprise was keyed up to the highest pitch. In the evening of the 28th of April news came that Borup and a college friend had been drowned within a few yards of shore while cruising in a canoe off Crescent Beach, Conn.

Regarding the death of Ross Marvin in the Arctic ocean Borup wrote,—"I am sure Marvin met death in a grand struggle in an attempt to heighten the glory of his country, battling alone against the forces of nature in the Arctic Wilds. I feel sure that he was proud of his finish and that a smile was on his lips to the last, and I can ask no more of fate than that it should grant me such a superb end." We feel that our young explorer's end was no less "superb" than that of Marvin, though it did not come amid the Arctic Wilds of which he was so fond and where he expected to win independent laurels, for he gave up his life in the vain attempt to save his companion, who was wounded, and there was a wistful smile on his dead face as if he had won out in the battle of life even though he had not accomplished the purposes on which he had set his heart.

At a meeting of the Executive Committee of the Board of Trustees of the Museum held on 15 May, 1912, it was resolved to postpone rather than to abandon the Expedition, and among other things — "That the Expedition be constituted a memorial of the young explorer who was so keenly interested in it and who was the mainspring of its present undertaking."

When the reorganization of the expedition was taken up last fall, Mr. MacMillan was made its sole leader. The accompanying statement gives subscribers and friends of the enterprise a condensed account of the reorganized expedition, its personnel and its revised and extended plans for scientific work in the Far North during the ensuing two years.

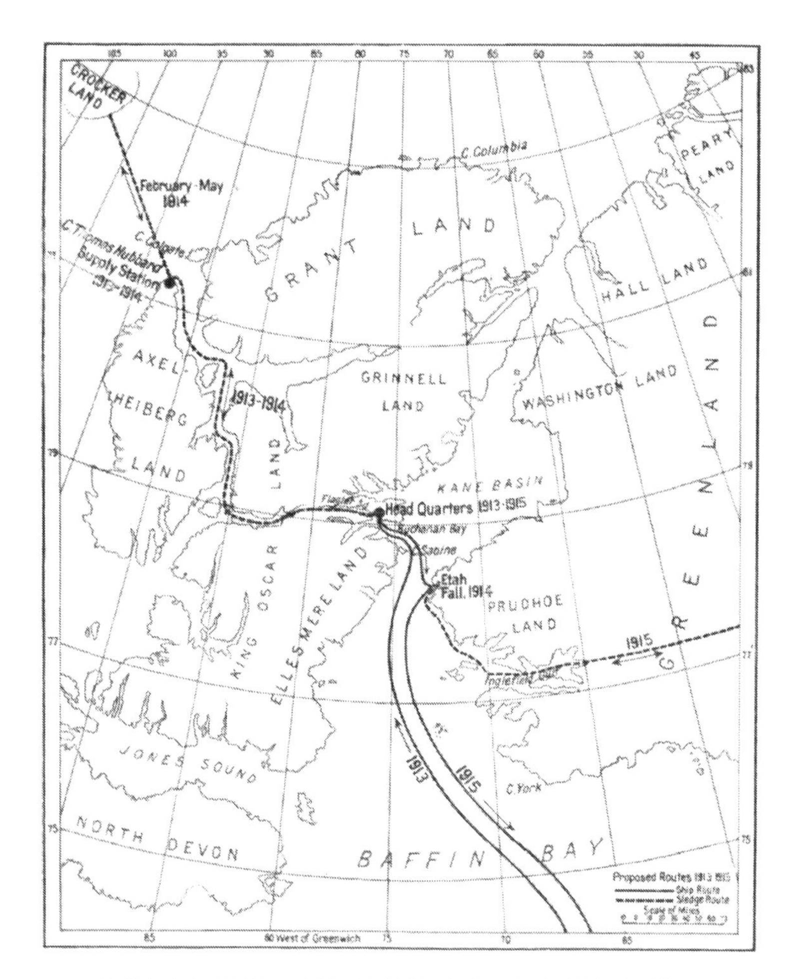

MAP SHOWING FIELD ROUTES OF THE CROCKER LAND EXPEDITION

OBJECTS

The objects of the Crocker Land Expedition are

1. To reach, map the coast line and explore Crocker Land, the mountainous tops of which were seen across the Polar sea by Rear Admiral Peary in 1906.
2. To search for other lands in the unexplored region west and southwest of Axel Heiberg Land and north of the Parry Islands.
3. To penetrate into the interior of Greenland at its widest part, between the 77th and 78th parallels of north latitude, studying meteorological and glaciological conditions on the summit of the great ice cap.
4. To study the geology, geography, glaciology, meteorology, terrestrial magnetism, electrical phenomena, seismology, zoölogy (both vertebrate and invertebrate), botany, oceanography, ethnology and archeology throughout the extensive region which is to be traversed, all of it lying above the 77th parallel.

SCIENTIFIC PROGRAMME

Geographical Work

1. The verification of the traditions of the Eskimos, of the reports of Captains McClure and Keenan, of the theories of Dr. R. A. Harris, the tidal expert, and of Rear Admiral Peary's discovery is left as the last great geographical problem of the North. Captain Richardson, in his work entitled "The Polar Regions," says: "The Eskimos of Point Barrow have a tradition, reported by Mr. Simpson, surgeon of the 'Plover,' [in 1832] of some of their tribe having been carried to the north on ice broken up in a southerly gale, and arriving, after many nights, at a hilly country inhabited by people like themselves, speaking the Eskimo language, by whom they were well received. After a long stay, one spring in which the ice remained without movement they returned without mishap to their own country and reported their adventures....An obscure indication of land to the north was actually perceived from the masthead of the 'Plover' when off Point Barrow."

In 1850 Captain McClure, when off the northern coast of Alaska, wrote in his journal that, judging from the character of the ice and a "light shady tint" in the sky, there must be land to the north of him.

To quote from an article in the National Geographic Magazine, 1894, "An Undiscovered Island off the Northern Coast of Alaska," by Marcus Baker: "It is often told that natives wintering between Harrison and Camden Bays have seen land to the north in the bright, clear days of spring. In the winter of 1886–1887 Uzharen, an enterprising Eskimo of Ootkearie, was very anxious for me to get some captain to take him the following summer, with his family, canoe and outfit, to the northeast as far as the ship went, and then he would try to find this mysterious land of which he had heard so much; but no one cared to bother with this venturesome Eskimo explorer."

The only report of land having been seen by civilized man in this vicinity was made by Capt. John Keenan of Troy, N. Y., in the seventies. He was at that time in command of the whaling bark "Stamboul," of New Bedford. Captain Keenan said that after taking several whales the weather became thick, and he stood to the north under easy sail and was busily engaged in trying out and stowing down the oil taken. When the fog cleared off land was distinctly seen to the north by him and all the men of his crew, but as he was not on a voyage of discovery and there were no whales in sight he was obliged to give the order to keep away to the south in search of them.

In June, 1904, Dr. R. A. Harris of the United States Coast and Geodetic Survey published in the *National Geographic Magazine* his reasons for believing that there must be a large body of undiscovered land or shallow water in the Polar regions. He based this theory upon the report that Siberian drift wood had been picked up in south Greenland, upon the observations of drifting Polar ice, upon the drift of the ship "Jeannette," upon numerous tidal observations made along the northern coast of Alaska and eastward. Knowing Harris's theory regarding such land, Peary, when he went westward along the northern shore of Grant Land in 1906, kept anxiously scanning the northwestern horizon for confirmation of it. On 30 June, 1906, as he states in his book "Nearest the Pole," he stood on the summit of Cape Thomas Hubbard 1400 feet above the level of the sea and saw distinctly "the snow-clad summits of the distant land in the northwest above the ice horizon." The explorer had seen the new land from 2000 feet up on Cape Colgate a few days before and, therefore, was enabled to estimate its distance as being about 130 miles from Cape Thomas Hubbard, in longitude 100° West and latitude 83° North. He named it Crocker Land in honor of the late Mr. George Crocker of the Peary Arctic Club.

2. Next in importance to establishing the extent of Crocker Land (or Islands) and delineating its coast line is the investigation of the region west and southwest of Cape Thomas Hubbard and north of the Parry Islands to determine what land there may be there.

3. Another question of supreme geographical importance is the location of the edge of the continental shelf. To quote Nansen ("North Polar Problems," 1907): "The determination of the extent of the continental shelf to the north of Axel Heiberg Land and Ellesmere Land would be a great achievement....The extent and shape of the Polar continental shelf, which means the real, continental mass, is the great feature of North Polar geography which is of much importance, geographically or geomorphologically, than the possible occurrence of unknown islands on this shelf."

4. The fourth task of importance in the geographical work is the delineation of the coast line along certain portions of Grant Land and Axel Heiberg Land.

5. The fifth great geographical problem to be attacked will be the nature of the Greenland ice cap from west front to summit in the widest part of the island, and this will include important studies on the rate of movement of continental ice masses and the collection of data on the phenomena of air currents over the mass of ice.

Oceanographical Work

The oceanographical work will consist of continuous tidal observations with a self-recording gage for two years at the permanent headquarters on Flagler Bay and of intermittent observations with other apparatus at Cape Thomas Hubbard, Crocker Land and elsewhere. The direction, strength and velocity of the ocean currents will be noted. Soundings will be made from boats and sledges.

Geological Work

Physiographic researches will be prosecuted continuously for the determination of the origin of the forms of land sculpture, of the regional cycle of erosion, of the special work of ice in summer and in winter. In dynamical geology, there will be a large field for investigation into the oscillation of the strand line; the extent, character and effect of igneous activity; the occurrence, nature and extent of dislocations, folds and faults. Research in physical geology will concern itself particularly with the problems presented by the accumulation, movement and dissipation of valley and continental glaciers. The phenomena of the meeting of the land ice with the sea ice will furnish an interesting item of study. Stratigraphic work will pertain to the differ-

entiation of the members of the Archean complex and to the ascertainment of the character, age, thickness and distribution of the sedimentary series, supplementing and extending the work of the Sverdrup and earlier expeditions, much of which was unavoidably fragmentary in character. Careful and comprehensive paleontological collections will be made in order to subdivide the sedimentaries into as detailed sections as may be practicable and to furnish ample material for comparative study. Collections will be made from the igneous rocks encountered and data obtained regarding their relations with and effect upon their associated rocks.

Geophysical Work

Observations in terrestrial magnetism along several lines will be carried on with an ample outfit for accurate work. Seismological disturbances will be recorded by means of a Wiechert pendulum. The electrical conditions of the upper air will be studied by means of the kites which are to be flown at high altitudes in connection with the weather station work. A high power wireless telegraph outfit is to be a part of the equipment, not only for the purpose of communicating weather reports to the Washington weather bureau and receiving the noon time signal from the Naval Observatory but also for the scientific study of many problems in the new science the solutions of which may possibly be favored by the conditions which obtain in the far North.

Chemical Work

As far as time permits, certain chemical work will be undertaken to observe the effect upon reaction of long continued temperatures of -60° to -70° F.—conditions that are difficult to maintain in the laboratory.

Meteorological Work

The meteorological work of the Expedition should be highly important in its bearing upon the theory of storms and their prediction. Continuous records for two years will be made by skilled observers on a full set of weather bureau instruments to be installed at the home headquarters on Flagler Bay, and daily reports will be transmitted by wireless to Washington.

Zoölogical Work

The land fauna will be studied, with special reference to the influence of climate, altitude and topography, and such skins and skeletons as are needed for scientific collections will be procured. The marine forms, both vertebrate and invertebrate, will be collected and studied and data secured with reference

to the influence of depth of water and currents. Particular attention will be paid to entomology, note being taken of the adaptation of insects to the short, warm seasons and their relations to other animals and to plants.

Botanical Work

The plants of the Arctic regions continue to offer a highly interesting field of study, particularly among gymnosperms and along the lines of ecology. It is planned, therefore, to devote a relatively large portion of the time assigned to botany to the associations of the plants and their succession and to make, if possible, a correlation of the character of the associations with the climatic and edaphic factors which constitute the habitats.

Ethnological and Archeological Work

The Smith Sound Eskimos were discovered by Sir John Ross about eighty years ago. They had thought that they were the only people in the world. The ruins which have been found dotting the region north of 79 degrees of north latitude indicate a much larger population than was found by Ross and still larger than exists to-day, for the Smith Sound tribe is dwindling. Contact with the whites has already seriously affected their life and customs, but they are still singing their weird native songs and reciting the traditions of their people. It is highly important to preserve these by means of phonographic records for future study and comparison.

DONALD B. MACMILLAN
Leader of the Crocker Land Expedition in winter costume

The proposed site of the main headquarters of the Crocker Land Expedition is amid ruins of the vanished people; while around the shores of Flagler Bay, through the mountains of Grinnell, Ellesmere and King Oscar Lands and along Eureka and Nansen Sounds are presented exceptional opportunities for archeological study.

Photographic Work

The camera will be used to supplement all branches of scientific work as outlined above. Therefore, an extensive outfit has been provided, including apparatus for color photography and the taking of moving pictures.

PERSONNEL

The leader of the party, DONALD BAXTER MACMILLAN (page 10) who was co-leader with George Borup of the expedition as originally planned, won his spurs in arctic work as one of Admiral Peary's trusted lieutenants during the successful quest of the North Pole in 1908–1909, traveling more than two thousand miles by dog team. Mr. MacMillan is a graduate, A. B. and Honorary A. M., of Bowdoin College. He has spent the three and one-half years since his return from the Arctic in studying, lecturing and traveling, always keeping in mind the present great enterprise as the goal of his ambition. The summers of 1910 and 1912 were spent in exploratory, ornithological and archeological work in and along the coast of Labrador. The year 1910–1911 and the past autumn and winter have been spent at Harvard University working in anthropology and practical astronomy. Mr. MacMillan is of Scotch ancestry and was born in Massachusetts thirty-seven years ago. He is a Fellow of the Royal Geographical Society of London and of the American Geographical Society of New York.

W. ELMER EKBLAW was born in Illinois and is thirty years old. His portrait shows him as he is, strong, sturdy, self-reliant, as befits his Scandinavian ancestry. He took his degrees A. B. and A. M. in course at the

W. ELMER EKBLAW

University of Illinois, specializing in geology, botany and ornithology. He is a valued instructor in geology at his alma mater. He has done much field work in that science and will have charge of the geological and botanical work of the expedition.

ENSIGN FITZHUGH GREEN, U. S. N., graduated at the Naval Academy, Annapolis, four years ago, when he was only twenty years of age. He was born in Missouri of old Colonial stock and received his appointment to the Academy from that State. Mr. Green's experience has been largely at sea, where his duties concerned navigation and all the complicated machinery

ENSIGN FITZHUGH GREEN, U. S. N.

of a battle ship. He was in command of a turret on the "Michigan" and has likewise done mapping of coast lines. He has been taking special studies during the past year in cartography, meteorology, seismology terrestrial magnetism and wireless telegraphy in Washington, D. C., and will have charge of these branches of the expedition work. His experience in the navy has already taught him how to command as well as to obey.

MAURICE C. TANQUARY was born in Illinois, though his parents and his grandparents for several generations were of New England birth and

MAURICE C. TANQUARY

training. He was graduated from the University of Illinois in the class of 1907 and continued his college work there to include the degree of Ph. D. He specialized in zoölogy, more particularly entomology, and he is well fitted to take charge of the zoölogical work of the party, including that on fishes and the mammals. Psychology was one of his important minor studies. He is thirty-one years old and is full of the life and vigor that will be invaluable to the party during the long hours of darkness in the second winter.

In addition to the strictly scientific staff of the expedition there will be a surgeon. The climate of the Arctic regions is so healthful for white men that there is not likely to be much for the doctor to do for the staff aside from the treatment of frost bite and the results of accidents, but the Eskimos furnish a fertile field for medical study and maintaining the health of the whole party through properly balancing the diet is of high importance. The surgeon has not been appointed yet though there have been many applications for the post.

The party includes THEODORE ALLEN an expert electrician detailed by the Navy Department to have the direct care of the wireless and other electrical outfit at headquarters, and his duties will include making the meteorological and seismological observations during the absence of Mr. Green. J. C. SMALL has been secured to serve as general mechanic and cook. The whole staff will seem nearly perfect, when a competent surgeon has been secured.

ITINERARY

Leave New York on the steam whaler "Diana" on or about 2 July, 1913. Call at Boston for pemmican and other supplies, at Sydney, N. S., for coal and sundry parts of the equipment. Touch at Battle Harbor, Hopedale and Okkak on the Labrador coast, and then head for the west coast of Greenland, skirting it to avoid ice. Cape York will be reached about 1 August, where dogs and dog drivers will be secured. If Flagler Bay, Ellesmere Land, is clear of ice by 20 August, the ship will push its way in and discharge its cargo and the expedition on the north shore. If ice conditions are unfavorable for the carrying out of this plan, the landing will be made at Payer Harbor, Pim Island. The ship will then return to New York.

The first task of the Expedition party will be the erection of the permanent headquarters. These will consist of a house for the white men of the Expedition party, another for the Eskimo dog drivers and their families and a third for the protection of the seismograph. The wireless telegraph too and other apparatus must be installed as soon as is practicable. While this work is going forward, a reconnaissance will be made to determine the best practicable route through the mountains of Grinnell, Ellesmere or King Oscar Land to Cape Thomas Hubbard. As soon as sledging can be undertaken, hunting parties will be sent into the hills to procure the winter's supply of fresh meat. During the late fall and in the moonlight of the Arctic night, supplies will be sledged to the secondary base, which is to be established on Cape Thomas Hubbard, caches of food being made along the route at intervals of about twenty miles.

CAPE THOMAS HUBBARD

The start for Crocker Land will be made with the coming of dawn in February, 1914. Unless unforeseen difficulties are encountered, the traverse of the sea ice will be made in about two weeks. Arriving at Crocker Land, the party will subdivide into three sections; one to go northward, one southward and one into the interior. The length of stay on

Crocker Land will depend upon the character of the ice encountered on the Polar sea and upon the amount of food it has been possible to carry.

Return to Cape Thomas Hubbard early in May. Spend about a month in exploring, mapping and studying the unvisited regions in the vicinity. Arrive at permanent headquarters about the middle of June, to spend the summer in scientific work there and in laying in supplies of walrus and seal meat for the following winter.

In the winter of 1914-1915 the foregoing programme will be repeated, except that a section of the party will be dispatched into the region southwest of Cape Thomas Hubbard in search of new land.

If all the predictions and observations are erroneous and "Crocker Land" does not exist, a journey to the summit of the Greenland ice cap will be undertaken in the summer of 1915.

At least one or two men will be at the permanent headquarters on Flagler Bay continuously for two years caring for apparatus and making observations on the weather, the seismograph and tides and receiving and transmitting messages with civilization through the wireless telegraph.

OUTFIT

The chief items of the outfit for the expedition are as follows: not less than three years' provisions for seven or eight white men, their helpers and their dogs; clothing; all the kinds of instruments needed for making the observations and records necessary to carry out the extensive scientific programme which has been outlined; moving picture and other photographic apparatus; a powerful motor boat and a large whale boat; salaries of some of the members of the scientific staff; a steamship to take the party northward in 1913 and another to bring it back in 1915.

COST

The original estimate was that not less than $50,000 would be needed for the necessary requirements of the expedition. This estimate was based upon a staff of only four or five white men and a less ambitious scientific programme. The expenses incident to halting the plans last year and to increasing the staff and extending the programme raised the cost of the expedition to $65,000, more or less, depending upon the expense of chartering ships for the two journeys.

Mr. MacMillan generously continues the original arrangement by which he serves as leader of the expedition without salary during the period of its absence from New York.

ORGANIZATION AND SUPPORT

On the proviso that sufficient funds were to be contributed from outside sources, the American Museum of Natural History started the enterprise by agreeing to appropriate $6,000 in money and assume its organization and management. The American Geographical Society made an appropriation of $6,000. Messrs. George Borup and D. B. MacMillan made themselves responsible for the raising of not less than $5,000 each. Yale University made an appropriation of $1,000. With this beginning, subscriptions were solicited with the result that by the end of April, 1912, the friends of Mr. Borup had responded with the payment of $10,608 and the friends of Mr. MacMillan with $6,454; a friend of Colgate University had contributed $1,000; sundry subscriptions in cash, supplies and equipment had been promised amounting to about $10,000, while tourists for the summer trip had been booked to the extent of $4,500, making a total of $45,562, so that the successful equipment and dispatch of the expedition were assured.

The postponement of the plans last year caused considerable loss through the change of tourists' plans and otherwise, but on reorganizing the expedition last winter the United States Navy Department detailed Ensign Fitzhugh Green for detached duty as a member of the scientific staff in charge of map work, seismology and electrical work and promised the loan of many valuable instruments; the American Geographical Society subscribed an additional sum of $3,000; several of the original subscribers increased their gifts and other friends of science and exploration added to the fund by about $4,000 in cash and $10,000 in supplies and apparatus. Then the University of Illinois came forward with a generous subscription which enabled the committees to add a trained zoölogist to the scientific staff and to provide for other needs of the expedition.

It is impracticable to insert in this edition of the prospectus a complete list of the institutions, federal departments, firms and individuals contributing to the equipment of the expedition, but such a list of acknowledgments will be published and distributed when the ship sails.

NEEDS

Only what was absolutely necessary to the proper subsistence and the scientific work of the expedition has been included in the outfit asked for through subscriptions, but the amusement side of existence during the long hours of darkness, particularly in the second winter, should also be provided for. Several publishing houses have made generous gifts of books and magazines and talking machines have been provided through the generosity of friends, but there are other respects in which this side of the life might be still better cared for and additional subscriptions would be welcomed.

PRESSURE RIDGE IN SEA ICE

CROCKER LAND EXPEDITION

TO THE

NORTH POLAR REGIONS

(GEORGE BORUP MEMORIAL)

STATEMENT TO CONTRIBUTORS

CROCKER LAND EXPEDITION—CHAIRMEN OF COMMITTEES AND STAFF

Upper Row, left to right—Henry Fairfield Osborn, Edmund Otis Hovey, Donald B. MacMillan
Lower Row, left to right—Harrison J. Hunt, Maurice C. Tanquary, W. Elmer Ekblaw, Fitzhugh Green, Jerome Lee Allen

Courtesy of American Press Association

CROCKER LAND EXPEDITION

(GEORGE BORUP MEMORIAL)

STATEMENT TO CONTRIBUTORS

The organizing institutions of the Crocker Land Expedition desire to make the following report:

The status of the Expedition up to 12 May was well set forth in the pamphlet entitled "Crocker Land Expedition to the North Polar Regions (George Borup Memorial)" which has been distributed to all contributors to the enterprise. Since the printing of that report, the Expedition has been fortunate enough to secure as surgeon, Dr. Harrison J. Hunt of Bangor, Maine. Dr. Hunt is a graduate of Bowdoin College of the Class of 1902 and has been practising his profession since his graduation from Bowdoin Medical College in 1905. In addition to his surgical and medical work, Dr. Hunt will make special studies in bacteriology.

The other members of the staff are Donald B. MacMillan, A.B., A.M., leader and ethnologist; W. Elmer Ekblaw, A.B., A.M., geologist and botanist; Fitzhugh Green, U.S.N., engineer and physicist, Maurice C. Tanquary, A.B., A.M., Ph.D., zoölogist, Jerome Lee Allen, wireless operator, and Jonathan C. Small, mechanic and general aid. Edwin S. Brooke, Jr., accompanied the ship as official photographer, and will return this fall.

When the "Diana" sailed from New York on 2 July, it was not practicable to publish a complete list of the friends who had given of their money or goods to the Expedition, but such a list has now been made up and is presented herewith.

The "Diana" took the major portion of the Expedition's equipment from New York, but she called at Boston on 4 and 5 July and took on board seven tons of pemmican, some boats and her outfit of chronometers and watches. Her next port of call was Sydney, N. S., where she loaded, among other things, twenty tons of dog biscuit and 337 tons of coal. She sailed out of the harbor of North Sydney on Saturday, 12 July, expecting to touch at Battle Harbor, Labrador, where but little time would be required for the loading of her thirty foot power boat, the "George Borup," and some caribou skins. Then she was to start on the long stretch of fifteen hundred miles to Cape York. The Strait of Belle Isle, however, contained much ice and the difficulties of navigation were increased by a dense fog, so that the heavily laden sealer made slow progress toward the north, and at one o'clock in the morning of the 17th she went hard and fast aground on the rocks off Barge Point, Labrador, near the provincial boundary between Labrador and Quebec and a few miles from the fishing station of Red Bay.

Fortunately, the sea was not rough and the old wooden vessel held together while signaled and telegraphed calls of distress brought fishing schooners and the government vessel "Stella Maris" alongside. The deck load of coal was jettisoned and other supplies from deck and hold were transferred to the fishing schooners and the "Stella Maris." The "Diana" was pulled off the rocks and the whole party went forward to Battle Harbor. Examination showed that the "Diana" was unfit to proceed to the Arctic regions and the steam sealer "Erik" of St. Johns was chartered to take her place. Some time was occupied in preparing the "Erik" for the journey, loading her with coal and getting insurance upon her, and it was necessary finally for the "Diana" to creep along down from Battle Harbor to St. Johns and make the transfer of cargo at the latter place. As soon as this was effected, Mr. MacMillan and his party started once more for the north, leaving St. Johns on Thursday, 31 July, and arriving at Battle Harbor at 8 a.m. on Sunday, 3 August. There they swung the "George Borup" on deck, took aboard the supplies which had been landed and left behind by the "Diana" and sailed on Monday, 4 August, for Cape York in latitude 76° N., on the west coast of Greenland, where they expect to make their first stop for the purpose of securing dog drivers and dog teams.

The reports received by Mr. MacMillan indicate an exceptionally open season in the far north, and he expects to find the Arctic waters freer from ice than they have been in twenty years or more, hence he hopes to be able to proceed without delay to Flagler Bay (Latitude 79° 10') and land his equipment by 20 August. If Flagler Bay is too full of ice, the landing will be made at Payer Harbor on Pim Island. When the landing has been effected, the "Erik" will return to St. Johns, arriving there about the middle of September.

Mr. MacMillan's letters express great satisfaction in the members of his staff and a feeling of certainty that they could not be improved upon for the work that lies before the party. Thus everything points to the highest success in the work mapped out for the Expedition.

AMERICAN MUSEUM OF NATURAL HISTORY
AMERICAN GEOGRAPHICAL SOCIETY
UNIVERSITY OF ILLINOIS

NEW YORK, 15 August, 1913

CASH CONTRIBUTIONS TO THE EXPEDITION

(GEORGE BORUP GUARANTEE)

W. W. Atterbury	$200 00
Herbert Austin	25 00
J. Sanford Barnes, Jr.	25 00
A. W. Beckman	10 00
Laurence C. Benét	75 00
Arthur C. Blagden	10 00
Henry G. Bryant	100 00
E. W. Clark	100 00
Class 1907 (Yale)	511 77
Harry E. Converse	40 00
Zenas Crane	500 00
W. R. Cross	25 00
Thomas de Witt Cuyler	500 00
James Lloyd Derby	15 00
Richard S. Dow	90 00
Grenville T. Emmet	100 00
H. Lloyd Folsom	25 00
Mrs. George B. French	250 00
George B. French	1,000 00
C. W. Gordon	100 00
Joseph C. Grew	25 00
Groton School	250 00
G. G. Grundy	20 00
M. H. Harrington	50 00
J. J. Higginson, Jr.	10 00
W. H. Hobbs	5 00
J. Frederick Hahn and C. T. Stewart	25 00
Lydig Hoyt	20 00
Thomas H. Hubbard	2,500 00
Mrs. Morris K. Jesup	500 00
Aymar Johnson	5 00
Fritz Kaufmann	1 00
L. D. Kellogg	1,000 00
C. H. Kelsey	50 00
Mrs. F. Larkin	$25 00
John Larkin	25 00
Charles F. Mathewson	50 00
George L. McAlpin	50 00
Ogden Mills	500 00
G. Frederick Norton	50 00
M. H. Neuwahl	5 00
New York Academy of Sciences	500 00
J. Donaldson Nichols	2 00
James C. Parrish	100 00
Samuel T. Peters	250 00
Lewis A. Platt	500 00
Frederick Potter	500 00
Samuel Rea	50 00
Edmund P. Rogers	5 00
John S. Rogers	5 00
Franklin D. Roosevelt	15 00
Theodore Roosevelt	10 00
Isaiah Scheeline	5 00
Jacob H. Schiff	250 00
Mortimer L. Schiff	250 00
T. L. Schurmeier	25 00
Henry Seligman	250 00
Isaac N. Seligman	250 00
Joseph L. Seligman	25 00
Robert P. Simpson	25 00
Harry Slutzger	10 00
B. Symonds	10 00
W. B. Thomas	200 00
Andrew G. Weeks	25 00
R. H. Williams	250 00
Yale University	1,000 00
	$13,379 77

(D. B. MacMILLAN GUARANTEE)

Edward O. Achorn	$5 00
Allen School	53 00
Anonymous	2 00
H. L. Bagley	25 00
S. Henry Baldwin	15 00
Harry Balfe	100 00
Bowdoin College	100 00
Frank E. Bradbury	10 00
D. L. Brainard	100 00
Arthur T. Brown	5 00
F. E. Clerk's Sunday School Class	$2 00
W. A. Clifford	1 00
Cogswell School	10 30
Frederick O. Conant	25 00
A. P. Cook	5 00
Charles Sumner Cook	10 00
Edward W. Cox	10 00
Zenas Crane	1,000 00
Philip Dana	25 00

John A. Devine	$1 00
John F. Eliot	5 00
Thomas J. Emery	10 00
Frederick A. Fisher	10 00
A. A. French	50 00
J. Arthur Furbish	5 00
Geo. M. Gray	100 00
Levi H. Greenwood	200 00
Clarence Hale	10 00
James C. Hamlen	10 00
W. H. Hastings	1 00
J. Everett Hicks	5 00
Rufus H. Hinckley	10 00
Henry Hornblower	200 00
Rayton E. Horton	25 00
Thomas H. Hubbard	2,500 00
William M. Ingraham	10 00
A. Marshall Jones	5 00
L. D. Kellogg	500 00
F. R. Kimball	15 00
John G. Knowlton	15 00
M. J. Look	37 50
J. W. MacDonald	10 00
Sumner T. McKnight	25 00
Charles F. Mathewson	50 00
George C. Menard	5 00
L. F. Mohr	5 00
Joseph E. Moore	10 00
Paul B. Morgan	100 00
Thomas F. Moses	$25 00
Franklin C. Payson	10 00
Henry S. Payson	10 00
N. Pierce	10 00
W. A. Powers	5 00
W. E. Preble	5 00
Prescott Club	60 00
W. A. Robinson	2 00
D. A. Sargent	10 00
C. A. H de Saulles	50 00
Otto C. Scales	1 00
George B. Sears	5 00
J. B. Sewall	5 00
Louis A. Shaw	100 00
Arthur L. Small	5 00
Edward Stamwood	10 00
W. D. Stockbridge	10 00
Charles A. Stone	200 00
John E. Thayer	1,000 00
S. C. Thayer	5 00
Townsend W. Thorndike	25 00
Wm. Underwood Company	613 10
Edwin S. Webster	200 00
Frank G. Webster	100 00
Hanson H. Webster	5 00
A. B. White	10 00
Harold S. White	1 00
Worcester Academy	300 00
	$8,214 90

(CASH CONTRIBUTIONS NOT DESIGNATED AS TO GUARANTEE)

Mrs. C. B. Alexander	$1,000 00
American Geographical Society	9,000 00
American Museum of Natural History	4,750 00
Colgate University	1,000 00
Henry Dodge Cooper	250 00
Zenas Crane	1,000 00
R. A. Harris	$50 00
Henry Fairfield Osborn	100 00
Peary Arctic Club	500 00
Robert E. Peary	500 00
University of Illinois	5,000 00
	$23,150 00

Tribune Contract (payable in 1913 and later) at least.......... $2,000 00

SUBSCRIPTIONS PAYABLE IN 1914–1915

American Museum of Natural History	$2,000 00
Zenas Crane	1,500 00
University of Illinois	5,000 00
	$8,500 00

ASSISTANCE IN OTHER WAYS THAN THROUGH GIFTS OF MONEY

FROM FEDERAL DEPARTMENTS

United States Navy Department—Detail for detached duty of Ensign Fitzhugh Green as engineer and physicist of the Expedition, and Jerome Lee Allen as electrician and wireless operator.

United States Navy Department—Storerooms at the Brooklyn Navy Yard and dockage for the "Diana."

United States Naval Observatory—Loan of instruments.

United States Hydrographic Office—Loan of a full survey outfit.

United States Department of Agriculture (through the Weather Bureau)—Loan of a full weather bureau outfit, together with kite equipment.

FROM INSTITUTIONS

Carnegie Institution (through Dr. L. A. Bauer)—Loan of full equipment for the study of terrestrial magnetism.

Georgetown University—Loan of Wiechert seismograph.

Harvard University (Division of Anthropology)—Measuring instruments.

Canadian Government—Free transmission through government wireless stations of daily weather reports and all scientific messages.

FROM FIRMS AND INDIVIDUALS

Abercrombie & Fitch Company.................. Pair of binoculars.
Alpha Portland Cement Company............. 25 bags of cement.
American Optical Company..................... Amber-colored snow glasses.
Atlantic Communication Co., New York, N. Y... Wireless telegraph equipment (loan).
Baldwin School................................ Gifts for Eskimo.
Dr. F. L. Banfield............................ 1 pair of ice creepers.
J. H. Bass & Company...... .... 1 pair of moccasins.
L. C. Bates (Paris Mfg. Co.),...................... Skiis.
John Bellman.......... .... Miscellaneous groceries.
Blish Milling Company......................... 4,500 pounds of flour and tins.
Borden's Condensed Milk Company.... .... 500 pounds of sweet chocolate.
H. D. Borup....... .... Adolph rifle with 200 cartridges to be presented to the Eskimo most useful to the Expedition and field glasses as second prize.
S. F. Bowser Company, Philadelphia, Pa.......... Special storage battery to supplement electrical outfit (loan).
Mrs. N. L. Britton..... .... 1 case of dried apples.
Carter, Carter & Meigs Company......... ... Medical supplies.
Central Oil & Gas Stove Company.............. Stoves, heaters, etc.
Colgate & Company .... Soaps, powders and perfumes.
Connecticut Telephone & Electric Co.......... Telephone equipment.

Direct Importing Company.....................100 pounds of tea.
Joseph Dixon Crucible Company................Pencils and erasers.
Mr. Dustin.... . ...Hunting knives.
Tom Fraser..... ....100 pounds of maple sugar.
General Electric Company, Schenectady, N. Y...Generating and transforming motors and full electricical outfit for the headquarters (loan).
O. C. Henk....................................Cigars, cigarettes, tobacco and pipes.
Hood Rubber Company..........................Rubber boots.
Edmund Otis Hovey............................American flag.
Howe Scale Company....................... ....Set of scales.
Knapp Company............................. ....200 reproductions of paintings.
Willard T. Libbey ....Pair of hunting knives.
Eli Lilly Company.............................Medical supplies.
M. J. Look....................................Rifle and 500 cartridges, 300 feet of belt lacing and gifts for Eskimo.
George Lowenstein ....Miscellaneous groceries.
H. W. McCandless..............................Detectors for wireless outfit.
National Dental Association...................Dental equipment.
Old Town Canoe Company......................Canoe, complete with sails, paddles, etc.
Mr. and Mrs. William Orme ....Gifts for Eskimo.
H. W. Phalen's Sons...........................Whip handles.
F. A. Patrick Company.........................20 yards Mackinaw cloth.
W. F. Patterson...............................Rubberoid roofing.
D. V. N. Person ....200 pounds of butterine.
Plymouth Cordage Company.............. .....50 pounds of rope.
Powell Chocolate Company....................200 pounds of candy, 30 pounds of breakfast cocoa.
Remington Typewriter Company................Remington typewriter with supply of ribbons (loan).
Riegel Sack Company...........................800 coal bags and 3,000 geological specimen bags.
John Russell Cutlery Company..... ....12 dozen knives.
Schieffelin & Company.........................1 box of bouillon.
Louis Agassiz Shaw ....1 mercurial barometer.
Singer Sewing Machine Company ....1 sewing machine complete.
Ellis Spear................................ .....1 automatic pistol.
Standard Oil Company..........................8,000 gallons of kerosene and 2,000 gallons of gasoline.
William S. Thomas.............................4 cases of canned beans.
U-All-No Chewing Gum Company ....Chewing gum.
United States Geological Survey..............Leather collecting bags (loan)
Victor Talking Machine Company..............7 victrolas and 200 records.
Waltham Watch Company......................Watches and chronometers (loan).
Samuel Ward Company....... ....Record books and field notebooks.
George Washington Coffee Refining Company..30 pounds of coffee.
William S. Watson.............................Miscellaneous groceries.
Alfred B. White......... ....Oilskin clothing.
Whitlock Cordage Company....................1 reel of twine.
Robert Wilson.................................Transit and telescope.
Wireless Specialty Apparatus Company.........Receiving wireless set and telephones.

WORCESTER SALT COMPANY........................16 barrels of salt.
YALE UNIVERSITY (Geological Department).........Yale flag.

The following firms gave special discounts on orders in consideration of the peculiar character of the Expedition, and as contributions toward defraying its cost:

ABERCROMBIE, FITCH & COMPANY
WM. AINSWORTH & SONS
AUSTIN NICHOLS & COMPANY
BURROUGHS, WELLCOME & COMPANY
CENTRAL SCIENTIFIC COMPANY
E. GERRY EMMONS
HENRY J. GREEN
LUMIÈRE ET JOUGLA
NATIONAL ENAMELING & STAMPING CO.
NEWMAN & GUARDIA
PARKER, BROTHERS & COMPANY
SPRATTS PATENT, LTD.
TOPPAN MANUFACTURING COMPANY
WM. UNDERWOOD COMPANY

LIST OF CONTRIBUTORS OF BOOKS TO THE EXPEDITION

J. A. ALLEN..1 volume.
AMERICAN GEOGRAPHICAL SOCIETY..........................8 volumes.
AMERICAN MUSEUM OF NATURAL HISTORY..................3 volumes.
ANONYMOUS..10 volumes.
JOSEPH BARRELL..1 pamphlet.
FREDERICK C. BEACH.................................Back volumes of the *Scientific American.*
CHARLES E. BESSEY......................................1 volume.
A. A. BRILL..4 volumes.
HENRY G. BRYANT..2 volumes.
BUFFALO SOCIETY OF NATURAL SCIENCES..................2 volumes.
CANADIAN GOVERNMENT..................................2 volumes.
STEPHEN R. CAPPS..1 volume.
CENTURY COMPANY.....................................Back volumes of the *Century Magazine* and 25 volumes.
MONTAGUE CHAMBERLIN ....1 volume.
FRANK M. CHAPMAN.......................................2 volumes.
C. B. CRAMPTON..1 volume.
EDWARD S. DANA ....1 pamphlet.
DOUBLEDAY, PAGE & COMPANY............................12 volumes.
D. G. ELLIOT...1 volume.
B. K. EMERSON ....1 pamphlet.
O. C. FARRINGTON..1 volume.
ANTHONY FIALA...2 volumes.
ARTHUR H. FOORD ....1 pamphlet.
GEOLOGICAL SURVEY OF CANADA......................... 3 volumes.
GINN & COMPANY...16 volumes.
GEORGE M. GRAY...$100 for purchases.
HARPER BROTHERS.......................................Back volumes of *Harper's Magazine* and 25 volumes.
A. G. HOGBOM ....2 pamphlets.
HENRY HOLT & COMPANY.................................24 volumes.
EDMUND OTIS HOVEY.......................................1 pamphlet.
LOUIS HURD ....1 volume.
L. D. KELLOGG...26 volumes (loan).

J. S. KELTIE	1 pamphlet.
LESLIE-JUDGE COMPANY	Back numbers of *Leslie's* and *Judge's* magazines.
LIFE PUBLISHING COMPANY	Back numbers of *Life*.
M. J. LOOK	$37.50 for purchases.
LONGMANS, GREEN & COMPANY	6 volumes.
E. H. MACKAY	9 volumes.
MARSHALL JONES COMPANY	10 volumes.
G. & C. MERRIAM COMPANY	1 volume.
A. G. NATHORST	1 pamphlet.
NATIONAL GEOGRAPHIC MAGAZINE	4 volumes.
F. R. ROWLEY	1 pamphlet.
SMITHSONIAN INSTITUTION	18 volumes.
DR. and MRS. SOULE	2 volumes.
GRANT SQUIERS	Collection of back numbers of magazines.
SMALL, MAYNARD & COMPANY	13 volumes.
STOKES & COMPANY	15 volumes.
UNITED STATES DEPARTMENT OF AGRICULTURE	4 volumes.
UNITED STATES GEOLOGICAL SURVEY	4 volumes.
C. B. WARNER	1 volume.
DR. WINTER	1 volume.

FINANCIAL STATEMENT

Receipts and subscriptions:			
Net cash received	$47,485 07		
Subscriptions payable in 1914 and 1915	8,500 00		
Tribune contract—at least	2,000 00		
Expenditures:			
Provisions		$13,716 88	
Scientific equipment		3,684 03	
Photographic outfit		2,070 83	
Books		368 90	
Camp equipment		3,731 46	
Coal		2,712 32	
Organization and administration		3,391 55	
Boat equipment		1,940 56	
Medical outfit		254 74	
Cancellations and storage		1,101 30	
Charter of S. S. "Diana" (2 months)		7,600 00	
Port charges, pilotage, etc		326 89	
Charter of S. S. "Erik" (2 months—made necessary by wreck of S. S. "Diana"		8,500 00	
Other expenses in connection with wreck of S. S. "Diana"		2,735 36	$52,134 82
Outstanding obligations			2,592 82
Salaries of staff to 1 October, 1916			8,222 50
Relief ship in 1916—estimated			11,000 00
Excess of expenditures and obligations over receipts and subscriptions	15,965 07		
	$73,950 14		$73,950 14

NOTE—There will be an expenditure for salvage on account of the wreck, through "general average," but the claim has not yet been adjusted and its amount cannot be even approximately stated.

IRVING PR
NEW YOR

Milton Keynes UK
Ingram Content Group UK Ltd.
UKHW020731271124
3180UKWH00049B/730